THE POWER OF
MATHEMATICS

MOSHÉ

FLATO

THE POWER OF

MATHEMATICS

McGraw-Hill, Inc.

New York St. Louis San Francisco Auckland Bogotá
Caracas Hamburg Lisbon London Madrid
Mexico Milan Montreal New Delhi Paris
San Juan São Paulo Singapore
Sydney Tokyo Toronto

English Language Edition

Translated by Maurice Robine
in collaboration with
The Language Service, Inc.
Poughkeepsie, New York

Typography by AB Typesetting
Poughkeepsie, New York

Library of Congress Cataloging-in-Publication Data

Flato, Moshé. 1937–
   [*Le Pouvoir des mathématiques*. English]
   The power of mathematics/Moshé Flato.
      p. cm. — (The McGraw-Hill *HORIZONS OF SCIENCE* series)
   Translation of: *Le Pouvoir des mathématiques*.
   Includes bibliographical references.
   ISBN 0-07-021258-9
   1. Mathematics. I. Title. II. Series.
   QA36.F5313 1992
510—dc20   91-45025

The original French language edition of this book
was published as *Le Pouvoir des mathématiques*,
copyright © 1990,
Hachette, Paris, France.
Questions de science series
Series editor, Dominique Lecourt

TABLE OF CONTENTS

INTRODUCTION

Could the content of mathematical theories and proofs be expounded in a natural language, in terms accessible to the greatest possible number of people? No, not in principle, reply the majority of mathematicians. Mathematics can be expounded only in mathematical terms: endlessly interactive, the generality of its abstractions has become such that it defies the irreducible metaphors of natural languages. Should the treatment of mathematics therefore have been omitted from the "Horizons of Science" series? The paradox goes a bit far when one realizes the power mathematics has gained, thanks to its abstraction, over the other sciences and, well beyond, over society as a whole. Moshé Flato, taking note of this paradox and of this power, does not deceive the reader: he refuses to take the easy way out of popularizing the subject by creating the illusion of understanding. The shortcut he follows is one of reflection on the power of mathematics in all its forms: arbitrary, legitimate and full of promise. Everyone reading it will be able to visualize the powers of invention and creation of mathematicians.

That mathematicians themselves wielded unequaled power was an idea formed well before such power began to be proven. Think of the early Greek mathematicians. They viewed mathematics as the key to explaining the world. Thus, according to Aristotle, who denigrated them, the Pythagoreans, in the 5th century B.C., "discovered that the modifications and the ratios of musical scales could be expressed numerically. Also, since everything else seemed to be modeled, by their whole nature, on numbers, and the numbers seemed to be the primary realities in all of nature, they supposed that the elements of the numbers were the elements of all things, and that the entire sky was a musical scale and a number." In fact, the intervals of octave, fifth and fourth could all be expressed in terms of simple numerical ratios. A surprising example of a phenomenon which had no apparent relation with mathematics and which, nevertheless, could be expressed in mathematical terms. The Pythagoreans thus had the grandiose vision of nature structured like music! But while that vision led them to indulge in many fantasies, which were soon held up to popular scorn, it had the advantage, however, of turning people's eyes toward the sky. Convinced that "the entire sky was a scale and a number," they were not content to listen to the inaudible music of the invisible celestial spheres where the stars were supposedly fixed; they positively opened the way to a mathematical astronomy. Conversely, it is also probable that they encouraged the study of acoustics at the same time, since one of their own, Archytas of Tarentum, became renowned for it late in the 5th century.

As for the actual accomplishments of the Pythagoreans in mathematics, our historical documentation is, to say the least, incomplete and often suspect. We are aware, however, that the irrationality of the diagonal of the square was known to them, either through exploitation of the consequences of the celebrated theorem that carries the name of Pythagoras, or through philosophical reflection on the idea of divisibility to infinity. It is also known that Archytas cleverly solved the problem of doubling the cube which long occupied the thoughts of the Greek mathematicians.

Plato advocated an ideal, or mathematical, astronomy over the astronomy of pure observation practiced at the time: "It is through the use of problems, as in geometry, that we study astronomy itself," he writes in *The Republic*. The mathematical genius of Eudoxus (ca. 408 – ca. 355 B.C.) would successfully explore that path. The same recommendation was made with respect to acoustics. And, as is known, Plato's *Timaeus* proposed, in mythical forms, a cosmology based on mathematics. While, like Empedocles (490–430 B.C.), he considered every natural substance to be composed of four simple bodies or elements (fire, air, water and earth), each of those simple bodies was, in fact, identified with a regular solid: fire with the tetrahedron (4 faces), air with the octahedron (8), water with the icosahedron (20), and earth with the cube (6). This is the first example of a geometrical theory of shape of the primordial bodies and the first attempt to reduce the changes taking place among them to mathematical formulas.

It is known, however, that the Greeks, though extraordinarily ingenious in mathematizing astronomy, never came up with the idea of mechanics and that a "technological block" ensued, the reasons for which have been disputed by historians for centuries. That situation is undoubtedly attributable to the weight of Aristotelian thought, which regarded this "sublunar" world as the stage for "generation and corruption" and assigned to physics only the task of qualitative elucidation of the constitution of bodies and their relations. Mathematics was thus reserved essentially for explanation of the circular motions of celestial bodies—perfect motions appropriate for beings conceived of as divine in nature. The movement of bodies on Earth did not seem mathematizable, on the other hand, since it was viewed, within the framework of that cosmology and of its metaphysical presuppositions, as a kind of change: as an imperfect transient "state" of the body, finding, under the impetus of an inner disposition, its "proper place," an "absolute" place where it would come to rest again in harmony with its essence (heavy bodies "down," light bodies "up").

It took Galileo, and then Descartes and others, to dismantle this qualitative "physics" and open the way to a mathematization of motion.

Thereafter, mathematics, together with modern physics, was destined to gain ever growing power. Yet one must not surrender to an exaggerated historical discontinuity: the scientific revolution of the 17th century was, in fact, favored by rediscovery of the work of Archimedes (287–212 B.C.) which, philosophically akin to that of Epicurus, forms with the latter "an already non-Aristotelian universe" (Michel Serres) not devoid of technological dynamism, at the heart of Greek civilization.

The Galileo affair has not yet delivered up all its secrets. A science historian named Alexandre Koyré devoted a vast amount of work to the subject, his scholarship advancing one essential thesis: it is not the sudden conversion to a supposed "experimental method" that is responsible for the birth of modern science, but a philosophic revolution, which leads one to think (again) of "the great book of nature" written in mathematical terms. But that implied a given conception and practice of mathematics.

This is very clearly evident when that mathematics is compared to what continued to prevail in China around the same time and kept it isolated from the birth of modern science, although the Chinese had for centuries made far-reaching strides in mathematical research. Bound to the political powers, mathematicians had been commissioned by the emperors to devise and reform the calendar; in the service of bureaucracy, they had notably developed an admirable body of knowledge useful in the practice of accounting and in the construction of grain silos, dikes and canals. Particularly flourishing under the Han dynasty (206 B.C.–220 A.D.) and Song dynasty (960–1279 A.D.), the Chinese mathematicians were thus essentially arithmeticians and algebrists. Joseph Needham has shown that in the 13th and 14th centuries the Chinese algebrists were in the very forefront, as their Arab counterparts had been in previous centuries, and like the Indian mathematicians who had invented trigonometry nearly one thousand years earlier.

What the Chinese undoubtedly lacked was a philosophical idea, the idea of a "legislator of the Universe" and, therefore, of the "laws of nature," and a method of mathematical reasoning: the abstract and systematic presentation of Euclidean geometry which "lent" itself to Galilean treatment, before later becoming an obstacle to the progress of mathematics and physics.

In any case, once a system of mechanics had come into being with Galileo, the power of mathematics never ceased to fascinate. The obstacles which Renaissance engineers had encountered seemed to be lifted: the precision of their calculations opened the doors to a technological world manifestly characterized by unprecedented practical efficiency and an abundance of goods. As early as 1648, Descartes was planning technical schools where masters knowing mathematics and physics could "enlighten" artisans in order to encourage and perfect their inventions.

With the *Principia mathematica philosophiae naturalis* [Mathematical principles of natural philosophy] (1687) of Isaac Newton, the power of mathematics became indisputable; it was exalted. It was discovered that the same fundamental law of attraction links the smallest and the largest bodies—atoms and stars—of the infinite Universe. But that law could be formulated and established only thanks to a transformation of mathematics itself. The immortal achievement of Newton consisted, in fact, in bringing mathematical entities together with physics and subjecting them to motion, no longer considering them in their "being," but in their "becoming" or in their "flux"; in short, of inventing differential calculus for physics, concurrently with his great rival Leibniz. Curves and figures were henceforth no longer constructed from geometric elements, nor cut out in space by the intersection of geometric bodies and planes, nor even conceived as relations of structures expressed directly by algebraic formulas. They were as generated or described by the motion in space of points and lines.

This commanding achievement established an idea of mathematics and its relationship with physics that persisted until the beginning of the present century. A work like that of Joseph-Louis Lagrange (1736–1813) marks a kind of apotheosis of that idea. Moshé Flato clearly shows how that "Newtonianism" has today come undone, how the unity of mathematics and physics has been restructured and reinforced, but on the basis of mathematics that has in turn been unified on the basis of "non-Newtonian" studies.

He also shows how that revision opens up new and exciting prospects for the mathematization of other disciplines like biology or, among the social sciences, economics. His plan is not unprecedented; but the terms in which his thought is couched appear almost totally original.

Let us take biology. The Newtonian success in physics did not fail to attract emulators—think of Buffon and his expressly Newtonian theory of "organic molecules"! But it very soon ran into the problem of "self-organization" of living organisms, a difficulty theorized in *Kritik der Urteilskraft* [Critique of Judgment] (1790) by Kant, who saw in the teleology inherent in living organisms an obstacle to any "physical" treatment. Auguste Comte and all the vitalistic biologists, examining the original relationships linking the whole with its parts in a living organism, proscribed the use of mathematics in that field. A short time later, however, the solitary work of Gregor Mendel (1822–1884) produced the first successful mathematization of a biological phenomenon, not only because he was bold enough to "apply" statistics to studying the segregation of characters in the course of generations of peas, the heredity of which he scrutinized, but because he caused a hitherto unnoticed biological structure to emerge through that calculation—one that would later be described as "one gene, one character." It is well known how Mendel's boldness ran into so much lack of understanding among his contemporaries that it took fifty years for the laws he had established to be rediscovered. The use he made of mathematics undoubtedly played a role in that lack of understanding, but perhaps, above all, the difficulty was strictly conceptual: Darwin himself, at the time he threw biology into turmoil by introducing the concept of "natural selection," conferred his authority on a theory of heredity, received from the 18th century, which failed to distinguish that question from the one of reproduction and which made the Mendelian "genetic" structure literally unthinkable!

Are we, in the years to come, going to see new areas of the biological sciences opening up to mathematization? One may think so, when considering, in particular, the present lineage of the research undertaken on animal morphology by Arcy Thompson (1860–1948) in his extraordinary book *On Growth and Form*, published in 1917. The discovery of "developmental genes," which shape the pattern of the organism, could well confirm the intuition he had of a biological structure linking evolution to mathematically measurable morphological variations. In any case, the reader will discover in this book that the best mathematicians prepare themselves, on their own, to participate in this adventure, even and especially if it means turning mathematics topsy-turvy in order to grasp those new subjects.

As for political economy, it has openly aspired to being mathematized for over a century and a half. Such an idea took hold, in fact, around the 1830s. The logician William Whewell (1794–1866) was already attempting in 1829 a translation of the principles of David Ricardo into algebraic symbols; but it was the French academician and philosopher Antoine-Augustin Cournot (1801–1877) who for the first time formulated the systematic program in his *Recherches sur les principes mathématiques de la théorie des richesses* [Research on the mathematical principles of the theory of wealth] (1838). Rejecting the questions raised by classicists Adam Smith and David Ricardo on the origin of value as "metaphysical" and reversing their reasoning, he took the market alone as object and turned to mathematics in order to confer the status of a science on his study. Cournot did not innovate in mathematics; he was content to apply mathematics, as he knew it (statistics and differential calculus), to an object, the "market," regarded by him in terms of mechanics, for explicitly normative purposes of establishing or restoring an ideal "balance." The science of trade (Léon Walras), the science of "transactions," economics has up to our times remained dominated by the way of thinking at work in its first models. Moshé Flato indicates under what conditions it could in the future renew its problematics and the role that mathematics could then play.

I now leave the reader to savor the pleasure of reading the book of a resolutely future-oriented mathematician of distinction, who, in discussing mathematics, expresses himself in clear, simple and forceful terms about our world. The reader will perhaps be surprised to lose familiar frames of reference: arithmetic, geometry, algebra, calculus, etc. But mathematics itself, after several aborted attempts in its history, has now erased them. Have your objects been hidden from you? Wouldn't that be the most manifest sign of the kinship indicated by Moshé Flato between mathematics and philosophy?

Dominique LECOURT

I

MATHEMATICAL
RESEARCH

CULTURAL AND SOCIAL POWERS

Whether an object of enthusiasm or simply of satisfaction, whether deplored or accepted with resignation, the power of mathematics is today an undeniable reality, solid and multifaceted. Not only do we speak of the efficiency of mathematics, but also of its domain—if not of its domination. Statistics holds sway over whole segments of social life: through opinion polls, it has conquered the world of politics, subjugating it, threatening political thought itself and distorting the democratic electoral process; voters are, in fact, insidiously persuaded to vote for the candidate leading in the polls, so as to be on the winning side, exactly the same way one plays favorites at the races, betting on the front runner, even while knowing the take will be paltry. The science of statistics was established a long time ago in the insurance world, while it was haunting the gaming tables—as it still does. Computer science, ideologically triumphal offshoot of 20th century mathematics, has contributed to the general restructuring of management practices, public and private, as well as to a veritable revolution in industrial and even artistic production.

This power arouses fascination and often exasperation. The widespread notion has, in fact, taken hold, not without reason, that a good knowledge of mathematics is the key and even ultimately the only real guarantee of social success. Parents are convinced of it and children are compelled, like it or not, to submit to the consequences of that conviction. In but a few years we have thus seen mathematics being used in Europe as almost the sole criterion of selective admissions to the best schools. The process is now too well known for me to dwell on it. It is to be noted, however, that this situation is not limited to secondary education, where it is most prominent, but in many disciplines touches upon the early years of university studies, reading even into the social sciences. Not only in economics, but also in sociology or psychology, students must assimilate a large dose of mathematics in the first couple of years of their curriculum, even though the usefulness of such learning might not be readily apparent to them at the time. It is especially regrettable that this knowledge, very often acquired with great hardship, may no longer be used to advantage in the rest of their studies, partly because in these fields the professors themselves do not have sufficient mastery of the subject.

It is to be added, finally, that this problem is particularly acute in France, where an ultraformalistic conception of mathematics—inherited from the great school of the 1940s and 1950s that acquired the collective pseudonym of Bourbaki—readily lends itself to that waste of knowledge. "Bourbaki," originally a student hoax, is an association made up almost exclusively of some of the prestigious Ecole Normale Supérieure's best alumni and alumnae, recruited by co-optation (and under 50 years of age), whose main purpose was to rewrite all of mathematics (the singular is preferred), ranging from the most general to the most specific, in a linear progression that became somewhat arborescent by virtue of circumstances. The published material of the mythical Nicolas Bourbaki was reputed to have come out of the University of Nancago, whose principal founders were originally professors at Nancy and Chicago. Their collective work was elaborated in the course of a very long process of maturation, including profound discussions during "retreats" in the Auvergne. Such an "encyclopedic" conception is not without risk to research, particularly in developing areas. Bourbaki seems, moreover, to be having more and more trouble following the present wealth of progress in mathematics and is today farther removed from what was its original object. Its publications are nonetheless often useful reference books, whose concise historical notes are accessible to the nonspecialist.

The primary risk of such a conception of mathematics is that it makes it impossible for a large number of pupils and students ever to gain a fair idea of what mathematical research theory can be. The picture they see is rigid and downright elitist. The situation also involves another risk, for it is not altogether certain that pupils described as "gifted in math" are really the ones with the greatest actual ability in this field. The approach taken in French education, from the second level of secondary school to its natural conclusion, the preparatory classes for the "grandes écoles" (selective State university colleges of France), is noted for its tendency to base admissions on mathematical prowess, which entails "cramming" for a large majority of students, a process that does not exactly stimulate the creative mind. How many of those children considered gifted, to the extent of being marked forever as "prodigies" by popular myth, are really only docile spirits whose powers of memory (for formulas and, at best, for concepts) are well exercised? That is serious, not only because it needlessly scares off many individuals who undoubtedly have other talents just as estimable, but also because it overly restrains the productivity of those who could have become true mathematicians.

EPISTEMOLOGICAL ILLUSIONS

But when the power of mathematics is evoked, not only do we speak of a social and cultural phenomenon, but also of the extraordinary sway that mathematics holds over the other sciences and, above all, over physics, in a relationship so close since the 17th century that, until the beginning of the present century, many scholars were both physicists and mathematicians or alternated in those pursuits. The wonderment at this power is as old as modern science. The greatest philosophers, starting with Descartes, tried to solve what, with the work of Galileo, already appeared to be an unfathomable enigma. It seems to me that the present evolution of mathematics demands reconsideration of the question with a fresh mind. It is not unreasonable to hope that we can see things more clearly today.

To embark upon such reflection, we must be convinced that mathematical research is indeed alive. Now, such conviction is manifestly not the case among the lay people, victims of mystifying ideas on this point. In fact, the prevailing opinion about mathematics, in contrast to other disciplines, whose advances or revolutions are envied, like the physical sciences or the life sciences, is that it leads a peaceful and definitely conservative life, that its moments of creative excitement

belong to the past and that there are "no more theorems left to discover."

Several factors contribute to this gross
misapprehension, the first being that research today
is identified with numerous teams at work in
laboratories fitted out with heavy equipment and
demanding costly investments. The image of "Big
Science," particle physics, nuclear physics or space
research dominates the popular mind. Now, there is
nothing like that in mathematics: by tradition the
people working in the field usually do so alone,
rarely in groups of more than two, and without
generally needing for their actual research work any
equipment other than paper, a pencil and a library.

It is to be noted, for example, that the supreme
award in the other disciplines, the Nobel Prize, is
commonly granted to several researchers for a joint
project or to heads of teams. In mathematics, on the
other hand, the Fields Medal, which is equivalent to
the Nobel Prize, has practically never been awarded
to other than individuals. And if there have been some
rare exceptions, the awards have not been presented
for team work, but rather have gone to two
mathematicians whose research was focused at a
given time on the same subject.

Another deceptive image adds to the misapprehension. Mathematicians are commonly regarded, in
fact, as simple virtuosos of calculation, whose "bril-
liance" has been measured ever since school days by

how fast they can calculate. In that case, it is hard to see what research might consist of, since it could simply be compared to an exercise of mental agility that modern computers, properly used, can often carry out far more rapidly and efficiently.

That these two images are actually caricatures without real foundation is clearly apparent when considering how today's mathematicians actually operate. First of all, one finds that there are more and more problems requiring the cooperation of several experts. This cooperation is sometimes voluntary and organized within a context of team work, but also very often it is simply a kind of convergence, each one contributing his or her own labor to the building, in sometimes widely divergent areas, until a brilliant architect creates an original structure out of it, as in the recent example of Mordell's conjecture, the proof of which won Gerd Faltings the Fields Medal in 1986. That conjecture involves an attenuated form of "Fermat's theorem," according to which the equation $x^n + y^n = z^n$ has no positive integer solutions for any integer n strictly greater than 2. (In the attenuated form the number of solutions of this type is finite; for $n = 2$ an infinity of solutions exists; the "theorem" is proven only for certain families of integers n.)

It is worth noting that mathematics has, since the turn of the century, undergone a process of inner specialization just as great as that which has affected the other sciences, particularly physics. That specialization has progressively blurred the classic distinctions known to all (even to the French Academy of Sciences) between algebra and calculus and geometry and mechanics. It goes without saying that, for sake of convenience, these terms are still used; but many now recognized areas of specialization straddle the old frontiers. To take just one example: today there is such a field as "algebraic geometry," originating from the work of Alexander Grothendieck, one of this century's greatest mathematicians. But many other fields emerging from the ruins of strongholds of the past could also be mentioned. In a controversial article which appeared several years ago, Jean Dieudonné, a walking encyclopedia and one of the founding fathers of Bourbaki, provided a classification of the different branches of modern mathematics worthy of the naturalist Carl von Linné, by grouping them in "divisions," in his view, of descending value.

This specialization, though rising out of the ruins, had the disadvantage, common in this type of situation, of shutting the door after each new area and rendering communication difficult, if not impossible, among the researchers engaged in those fields. Now, an increasing number of new problems is cropping up at the interface of these fields strictly delimited just of late. On any particular point, we thus see cooperation being organized among experts in algebra, calculus and topology. And very important conjectures, which are at the heart of mathematical progress, often come out of such collaboration. To take one good example, the work on the "index formula" important in calculus can be mentioned, standing in fact at the frontier between calculus and topology. That work has been successfully conducted through different team efforts. In reality, then, there have been more and more examples of such collective efforts, even though the bulk of the work in mathematics still consists of individual research.

As for calculations, one thing is certain. It is assuredly impossible to be a mathematician without making calculations. Mathematicians, in essence and by destination, cannot avoid them. But the meaning of this word must be made very explicit: mathematicians do not make only numerical or algebraic calculations, equations or, in short, calculations of the type made by engineers; they calculate with abstract, general objects. Those calculations are, furthermore, ever more abstract and concern ever more general objects: leaving aside rules of calculation and objects nurturing the usual numerical calculations, they make it possible to establish new relations between objects in the form of new theorems (at that stage, in fact, calculation merges with reflection). This growing abstraction and generalization signals a powerful unifying movement whose philosophical consequences should be carefully weighed.

There is then no reason to deny the existence of genuine mathematical research; it follows its own path and, at present, is very strong. Not only can it be shown that the number of publications is increasing very significantly, but it can be said that mathematics is basically experiencing considerable disruptions. The appearance of such disruptions, starting in the 1960s, prompted the restructuring of mathematics teaching in secondary schools and then at the primary school level, under the adopted name of "new math," which has gone through numerous agonies of youth (a subject to which we shall return later). It had become necessary to train the minds of pupils, from the earliest age, in the mental gymnastics required by mathematical research and useful in many other areas. That this is possible is underscored by the emergence in some countries, where the educational system permits, of brilliant students who sometimes obtain their doctorates even before they come of legal age: youth and the absence of maturity are not handicaps for some forms of mathematical thinking.

THE REALITY OF RESEARCH

Mathematical research, to characterize it in general terms, consists both in trying to discover new relations between known mathematical objects and in imagining problematic situations where the known objects no longer suffice to formulate the problems. The first aspect of such research obviously involves work in mathematics that is self-contained, prompted by inherent difficulties in forging ahead. That this aspect of mathematics is of the utmost importance can be shown by numerous examples. The most famous in our time is undoubtedly the example concerning the set theory, which gave rise to the formulation and solution of "Russell's paradox" (see below) in the celebrated *Principia mathematica* of Bertrand Russell, written early in the century in collaboration with Alfred North Whitehead. An inner need for precision represents the fundamental reason for such research.

But mathematical research can also be developed under the impetus of other disciplines, and here, in this second aspect, physics can be said to have played and to still play the decisive role, even though one might think and hope that other disciplines may contribute to it in the near future. Just think of the mathematical work of Newton: it is known that the initial question was not at all one intrinsic to the field of mathematics, but rather a problem of physics. Trying to formulate the laws of physics correctly, Newton realized that he could do so only by developing a new mathematical instrument: that is how he came to invent differential calculus. In general, classical physics has controlled the formulation of numerous mathematical problems. For lack of an existing formalism for solving one new problem or another arising in physics, a constant enrichment and development of mathematics was necessary. The situation is, of course, still often encountered today.

But what is less well known is that in the last few years, ten or twelve at most, a new way of mathematical thinking has emerged, which somehow reverses the traditional process and has given rise, before our very eyes, to what might be called—by symmetry and contrast with classical "mathematical physics"—"physical mathematics." According to this way of thinking, the formalism and methods of physics help us put problems to the test and to formulate strictly mathematical theories. This new development is, one might say, destined to have a great future and to help erase the sharp dividing line that has existed since the end of the last century between physics and mathematics. Examples are now numerous of how a mathematical problem can be attacked by means of a formalism borrowed from physics, or even of how physical formulas are used to prove mathematical theorems. Let us take just one of those examples, without, of course, going into technical detail: the field theory so important in physics was established by resorting to mathematics, which made it possible, for example, to determine the movement of one particle or another in given conditions, as well as many other things. We are dealing here with the traditional path which, in modern science, leads via mathematics from one branch of physics to another.

But here we are today resorting to field theory to solve certain questions relevant to mathematics alone, like those of knot theory (topological classification of knots and the invariants associated with them), or drawing inspiration from it to advance mathematical theories like that of operator algebra. The opposite path is then taken: from one branch of mathematics to another via physics, even if it involves, of course, a now highly mathematized physics. The latest Fields Medals, starting with that awarded Alain Connes in 1982, clearly illustrate this new trend.

Another way of presenting the same situation in simple terms: it might be said that mathematicians are always divided into two very different camps, corresponding to two distinct ways of thinking in mathematics. The first camp proceeds on the basis of analytical thinking, which concerns the continuum and maintains a privileged relationship with mechanics. Let us say it embraces "Newtonian" mathematicians who, on the basis of differential equations, can always imagine a mechanical model underlying their way of thinking. The mathematicians of the second camp, however, are not concerned with the continuum, but with the discrete, as is the case, notably, in number theory. Should they be described as "Pythagorean"? I leave this decision to the science historians and the few philosophers who might be interested. In any case, the thinking of these mathematicians is certainly not propped up by some mathematical model; it appears thus to be "more abstract" and demands new insight, when imagining physical models which might correspond to their conceptual elaborations.

These two schools of thought have, of course, always maintained close relations historically. One does not have to imagine a tradition split into two completely separate "lines," a modern version of the traditional division between algebra and calculus, or the division (dating back to antiquity) between arithmetic and geometry. Therefore, the partition I am proposing hardly seems questionable to me: two profound and persistent trends in mathematical thought are involved; and it seems to me that throughout history mathematicians have displayed particular talent in the register of the continuum as well as in the register of the discrete. Now, we are seeing these two schools of thought drawing closer and closer together.

THE UNITY OF MATHEMATICS

It may indeed be convenient, for didactic reasons, to divide mathematics into major areas which can then in turn be subdivided. But it would be erroneous to regard the boundaries between them as quasi-inviolable, as was the boundary that separated the Germanic from the Romance dialects in the heart of Western Europe, virtually unchanged since the days of Charlemagne. Quite the contrary, a majority of the most fertile and natural notions and developments in mathematics have been increasingly those which straddle different areas, in which they have sunk roots and from which they have drawn sustenance, integrating those areas into a sort of multidisciplinary sphere intrinsic to mathematics which goes far beyond what is encountered in other sciences and even the arts. The group theory, with all the transformations it has undergone and is still experiencing, is a typical example of such reconciliation of the different schools of thought at work. Emanating from "discrete" thinking, the internal logic of the development of mathematical thinking has, in fact, compelled it to sink ever deeper roots into the register of the continuum and then develop in the same breath on both levels, each nurturing the other.

Without for the moment going into detail, it is known that the general group theory, as we understand it today, originated with Evariste Galois (1811–1832), that last precocious genius of French mathematics, whose brilliant work was brought to light in 1843 by Joseph Liouville (1809–1882). Augustin-Louis Cauchy (1789–1857) and Joseph-Louis Lagrange (1736–1813) had studied permutation groups, and Camille Jordan (1838–1922) had presented a formalized version of it in his important work *Théorie des substitutions et des équations algébriques* [Theory of algebraic equations and substitutions] (1870). But the decisive event, in my opinion, was the elaboration of the continuous group theory by Norwegian mathematician Marius Sophus Lie (1842–1899). That theory has since borne his name and today we speak of Lie groups.

A Lie group is a transformation group of a given space, usually realizable as a matrix group, the elements of which depend on several parameters and for which a meaning can be given to an infinitesimal action, the limit of which is called Lie algebra.

This idea is now planted, with its different generalizations and realizations (representations), at the heart of mathematics. It embraces differential geometry, algebra, calculus, measure theory, algebraic topology, arithmetic, etc., and is one of the bases of mathematical physics. In short, it tends to unify domains and subdomains which have in the past century and a half been developed separately in extraordinary fashion.

We can thus see in what sense it is certainly not improper to speak of the progress of knowledge in contemporary mathematical research. This progress, which the simple formalistic, academic presentation of results and formulas conceals from the view of persons not participating in it, actually seems much clearer and more profound on the conceptual level than in disciplines like molecular biology, whose praises are nevertheless generally sung, even though it has experienced relative theoretical stagnation since the passing of the great era of scientists like Monod, Jacob and Lwoff.

This unification is so powerful that "mathematics" can be referred to in the singular, as I have already been doing all along. That is, moreover, what Bourbaki was alluding to on titling his treatise *Eléments de mathématique* (Elementary mathematic), and the order in which the work is arranged (linearly, each part depending on what comes before) underscores it. The unity is, moreover, even stronger than the Bourbakist approach suggests, for in modern mathematics the cross-links between different areas are constantly growing, with a multiplying effect on the resulting progress—somewhat like cross interests in the field of finance often being the source of economic progress.

PROGRESS IN MATHEMATICS

But progress in mathematics presents itself in a particular light in that it occurs according to two very different modalities. The first consists in solving classical problems, whatever their otherwise "internal" or "external" origin: when mathematicians gain a foothold in mathematics, they always run into one conjecture or another advanced in the past on the basis of a classical theory and still awaiting solution. They can take a stab at it. Not all those conjectures have the same interest as Fermat's theorem, but there is often much to be learned from such attempts. It is said that such mathematicians are *problem solvers*; they solve problems formulated on the basis of theories established by others and then left in abeyance.

There is no reason to neglect or disdain such activity. But another way of making progress in mathematics does exist. It consists in building new theories (devised by so-called *theory makers*). In view of the prestige attached to such activity, the utmost vigilance is indicated, for there are many shams practiced and a lot of counterfeit currency is put into circulation! The usual type of forgery consists in snatching up an existing theory and weakening it, for example, by removing an axiom. That is known to have happened in France in the case of group theory: researchers usurped temporary glory by settling for such depreciation and passing it off as something new! Another very different case, not involving a genuine innovation either, is that of category theory. It provides mathematics with a very precise and efficient language, but by itself has produced very few important mathematical results. By contrast, we see new theories taking shape, which are not presented as insipid versions of existing theories and turn out to be very prolific, because by delving into certain phenomena, they suggest to us the study of hitherto unknown structures.

Consider the recent example of microfunction theory, which comes to us from the Kyoto school of Japan, directed by Professor Sato or perhaps the better known and older example, (yet along the same lines) of Laurent Schwartz's distribution theory, will speak more to the reader. When it was formulated, there was no question of an artificial, degenerescent and superfluous refinement of some pre-existing theory, nor of tidying up the mathematical language, but of the definition of structures that were enriching mathematics and opening up the opportunity to think about and solve new problems—although this theory might be encountered in more or less embryonic form among theoretical physicists like Paul Dirac and experts on partial derivative equations like Sobolev, with the hard topological problems it raises being solved by Alexander Grothendieck in his thesis.

The extreme case of what I would call a *theory maker* appears to be represented by René Thom. If he arouses the passions of controversy, it is not just because he has the courage to express his opinions forcefully or vividly or because his language abounds in alluring expressions of poetic elegance (with allusions to butterflies, ruffles and magic spells—or catastrophe); it is because he is a kind of *theory prophet*: in his youth as a mathematician, he had the genius to foresee in broad outline the development of what is called "differential topology" and to establish its mathematical bases. (This is a field devoted to a comprehensive study of differentiable varieties, absolutely not reducible to local study, which is at the root of differential geometry.) He was able to open up a whole area of contemporary mathematics and encompass it all within his sight. But his genius, which was not applied to the solution of classical problems, also always disdained the technical solution of the very problems that his own theory unveiled.

It is upsetting that his public and media recognition was acquired too late, well after his extraordinarily productive period of the 1950s and for reasons which have nothing to do with his mathematical work. Thom himself readily acknowledges it: the catastrophe theory is not a mathematical theory, nor is it a biological or economic theory; it is an original way of thinking which he is convinced can be productive in different areas of research where the prevailing positivism has crushed inventiveness. That being said and accepted, with reservations, the major importance of Thom's mathematical work must not be forgotten.

When genuine progress occurs in mathematics, it is, in effect, a new way of thinking that is beginning, a new way of seeing things. Now, one might say that such productive changes have come about in the past thirty years and a series of unprecedented problems is thus open to research.

Think, for example, of the work of Alexander Grothendieck, continued by Pierre Deligne, and of Gerd Faltings, Alain Connes and Vaughan Jones; or of the results obtained by Michel Freedman in 1982 and by Simon Donaldson in 1983 (the Fields Medal was awarded to them in 1986) on four-dimensional space, fruit of new ideas worked out in common by topologists and physicists.

Most of these examples illustrate moreover a new phenomenon, which underscores the unity of mathematics: how, with the ulterior motive (usually unavowed and unachieved) of being a *problem solver* (in the case of a problem whose solution has escaped generations of mathematicians), one is inescapably led to become a *theory maker.*

PHYSICS AND MATHEMATICS

There is perhaps some point in comparing the pace of progress in physics with that which mathematics has experienced, in order to cast all necessary light on the specificity of the mathematical approach as such.

If we consider, in fact, the very spectacular developments in physics that occurred at the turn of the century, without going into the historical origin of concepts and theories, but rather confining ourselves to the results obtained, it can be said that, when going from classical mechanics to quantum mechanics, a parameter (Planck's constant h) is added. And it can also be said that classical mechanics is "distorted" by adding that parameter, which is a constant, the moment that parameter is different from zero. It is experience which teaches us to establish its value: at that point we enter the field of quantum mechanics. If, on the other hand, zero value is ascribed to that parameter, we are back in classical mechanics. The same remarks could assuredly be made about the special theory of relativity: if the parameter of the speed of light c is added to classical mechanics, the special theory of relativity is obtained. If, starting from the latter, it is said that the speed of light is infinite and then that $1/c$ is equal to zero, we are back to nonrelativistic classical mechanics.

It is evident from these two major examples how it can be asserted—after the fact—that the old formalism actually represented an approximation of the new. Thus, it appears that if the new formalism proves applicable for some dimensions (quantum mechanics for very small distances, the theory of relativity for very high speeds), it does not entirely cancel out the old, as has too often been fallaciously stated. It is, in fact, clear that the old theories remain fully valid within their own orders of magnitude. This has commonly led to the use of two types of theories in physics: if, for example, the speed of a given system is too great compared to the speed of light, the work is continued with the laws of classical mechanics; if, on the other hand, it approaches the speed of light, it is necessary to introduce relativistic corrections. By no means can it then be said that the classical theory is being "abandoned"; we simply go beyond it.

In mathematics, matters proceed somewhat differently. The new objects treated give rise to new mathematical needs. And the theorems then established often appear "stronger," more powerful, because they generalize the old ones. The old theorems, as in physics, are thus thought of as special cases of the new. Let us further explain: mathematical progress is progress toward a growing generality and abstraction of theorems. Here, then, by addition of a supplementary hypothesis, we automatically encounter the previously known theorem again. It could almost be said that the path of progress is symmetrical with the progress observed in physics.

Such generalization can, of course, open up a field of objects for which no old theorems exist— indeed the clearest measure of its productivity—aside from cases where one can, with some added hypotheses, fall back on already established theorems. But in no case can it be said, here either, that the old mathematics is being abandoned to make up new mathematics from it, with all due deference to the philosophers who have ventured blindly into this terrain.

If we are to grasp the meaning of the possibilities afforded research today in both areas, this comparison needs to be completed. After the golden age at the beginning of this century, which witnessed the successive appearance of the special and then the general theory of relativity and, finally, the even more revolutionary theory of quantum mechanics, it seems that physics has not experienced any further major theoretical revolution.

Many efforts have, of course, been made to wed these theories in what is called the "quantum field theory," and it must be granted that they have been essentially successful in the case of "quantum electrodynamics," which achieves a complete linkage between special relativity and quantum mechanics by treating (with a physicist's precision) the fundamental problem of interactions between the unit of light (the photon) and the electron. That theory won Julian Schwinger, Richard Feynman and Sin-itiro Tomonaga the Nobel Prize in 1965; each, in his own style, had independently developed it in the 1940s. It must not be overlooked, however, that, in spite of the extraordinary degree of precision of the predictions this theory makes possible, the rules of calculation employed here are not entirely satisfactory from the mathematical standpoint. Above all, as far as the "quantification of gravitational fields" is concerned, this linkage has still not been achieved. Every hope has so far been disappointed, in spite of the different formal approaches taken toward this problem.

We should not be surprised by this period of quiescence, however distressing we might regard it: looking at its history, we realize in fact that the progress of physics has always been marked by sharp discontinuities.

It seems to me that this kind of discontinuity is not accidental, but is due rather to an intrinsic characteristic of research in this field. Physics is a science which makes progress only when it encounters a paradox or comes upon a phenomenon that cannot be explained by the concepts and theories available at the time of its discovery. An inexplicable new phenomenon forces physicists to think, creating on such occasion a wave of panic in the profession, as we have seen several times since the last century; physicists are then put on notice to build a new theoretical framework to explain it; that is the typical course taken by the great advances in physics.

The case of quantum mechanics is once again very illuminating. It was the enigma constituted by "blackbody" radiation, strictly inexplicable in terms of classical physics, that compelled scientists to probe farther. The blackbody is a model which, in classical physics, radiates an infinite energy, for it is composed of an infinity of oscillators radiating the same energy; it is the introduction of Planck's constant (h) that enables the quantity of energy radiated by that model to be rendered finite. It is known that Max Planck was sorely troubled by this obligation: it was a profound revolution in thinking that his discovery put on the agenda, leading to developments in "wave" mechanics with the work of Louis de Broglie and to Niels Bohr's "planetary" model of the atom newly proposed by Ernest Rutherford and resulting in the work on quantum mechanics proper by Werner Heisenberg, Erwin Schrödinger and then Paul Dirac toward the end of the 1920s.

Now, it must be admitted that, even if they do exist, we are unable today to verify "paradoxical" phenomena of such amplitude. Although experimental results are being accumulated, new particles are constantly being discovered, and progress is being made on the phenomenological plane, there is still no solid and complete theory to explain and correlate all these phenomena. For example, for strong nuclear interactions, there is no coherent theory at all. In such a situation, no solid basis exists for detecting the presence or absence of a paradox.

In other areas, like solid-state physics, there are many remarkable experimental discoveries for which no theoretical model yet exists, like that of superconductivity (almost zero electric resistance) at "high" temperature ("higher" than the melting point of nitrogen), which won a recent Nobel Prize for two researchers of a small IBM laboratory in Zurich working on a material developed by a French chemist; the classical model of super-conductivity, which won John Bardeen, Leon Cooper and John Robert Schrieffer the Nobel Prize in 1972, applies to temperatures near absolute zero. But these discoveries do not seem to raise questions of principle. The theoretical problem of cold fusion (fusion—as in the Sun, but at room temperature—of light or heavy hydrogen nuclei in a palladium system), whose highly controversial experimental "discovery" made headlines a couple of years ago, can likewise be treated by models founded on the quantum theory, as the founder of quantum electrodynamics, Julian Schwinger, recently demonstrated.

Exciting questions are being asked about unification of the forces of interaction, and unexplained phenomena like the genesis of particles and the invisibility of their supposed constituents called quarks are being pondered, but in no case is quantum theory the subject of a contradiction. In fact, it is even constantly being confirmed experimentally.

And yet we must wait for such a contradiction in order to determine the new direction to be explored in the future. Until physicists come face to face with a paradox, we can expect a multiplicity of the most sophisticated mathematical models, as is the case today with the "chord" and "superchord" theory (whose relation to the real is dubious, to say the least) to remain speculative exercises. The day the contradiction occurs, we shall know it is necessary to generalize and at what level it would be desirable to do so. Of course, the development of certain mathematical models will perhaps suggest ideas for going farther in physics.

In mathematics the pace of progress is entirely different. In fact, that progress appears much more continuous, for there is no paradox to be awaited before guiding the action of generalization. Assuming that the progress of all natural sciences potentially responds, on the whole, to the characteristics just enunciated with regard to physics, it may be said that mathematics is a science "apart." Many thinkers, mathematicians, philosophers, and epistemiologists have tried to characterize this manifest peculiarity. It seems to me that it is through the continuity of this development that it can be most adequately grasped.

The history of set theory proves highly instructive in this connection. It seems to contradict the thesis defended here, since it elicits a very famous paradox, Russell's paradox, already alluded to, which compelled theoreticians to axiomatize the set concept and to learn its limitations: "the set of all sets" is, in fact, a contradictory concept.

But here the paradox does not lead to any change in research development. It is true that it was necessary to invent new axiomatics in order to prove certain theorems more precisely, but that gain in precision has not changed anything essential nor contributed anything to the content of said set theory, which has continued to be used and enriched as before.

Generally speaking, in mathematics no conceptual revolution has ever been known which has, as in physics, led to considering the former theoretical state of a given field as an approximation of the new thinking. In short, in mathematics we always generalize and increasingly innovate, while in physics areas of applicability of concepts are changed by "bending" the old theories in the right direction.

Forcing the issue slightly and, at the risk of offending many prejudices on both sides, I shall say that mathematical thought—which, lest we forget, existed long before the establishment of physics as a science, in the present sense of the term—appears to be closer to philosophical thought than to that of the natural sciences, and the process of mathematical creation has similarities to that of artistic creation. But, to understand it, we must now enter more directly into the realm of the bond between mathematics and the other sciences.

II

MATHEMATICS AND
THE OTHER SCIENCES

MATHEMATICS AND
MATHEMATICAL LOGIC

The undeniable abstraction of mathematical objects
has to be appreciated at its true worth: what is
involved is an active and fruitful abstraction,
sometimes endowed with an aesthetic character to
which no reasonable mind can remain insensitive.
But that fruitfulness is never guaranteed in advance,
and the concrete outcome of a "pure" mathematical
creation can take a long time coming; it can also
come by surprise, when not expected or no longer
expected.

The first mistake to be avoided, if we are to
grasp that inner vitality of mathematical thought, its
audacity and its beauty, consists in seeing in
mathematical objects just so many fruits of pure and
simple logical processes. Many philosophers and
some mathematicians of renown have sown
confusion on this point since the turn of the century.
Let us be quite clear about this: a philosophical
logic used to exist, originally associated with the

metaphysics of Aristotle, and then adopted and fundamentally modified by great thinkers like Kant or Hegel. A mathematical logic has since developed which cannot be equated with the philosophical type. This new logic is a branch of mathematics; it is thus without authority to provide a so-called "foundation" for mathematical abstractions; nor can it serve as a hidden mainspring for their sequences. It is just as futile and unreal to seek in this logic some *a priori* assurance of "applicability" to the physical world. On the other hand, it is of the utmost interest to ponder the particular course of the development of logic and the ties it is able to maintain with the other branches of mathematics— therefore to detect the interface problems brought to light by the contacts it is able to maintain with the rest of mathematics. The question proves all the more subtle as logic occupies a reflexive position in relation to mathematical development; it appears second to the actual creative work in mathematics which, for the most part, always precedes it.

As for its development, one can say that, stemming from the work of Bertrand Russell, Gottlob Frege, Ludwig Wittgenstein and others, it provided the setting for two major intellectual revolutions which are, in fact, linked, with which the names of Kurt Gödel and Paul Cohen are associated.

The 20th century began in mathematics with the very ambitious program announced to his peers by the mathematician from the University of Göttingen, David Hilbert (1862–1943), at the Second International Congress of Mathematicians held in Paris in the summer of 1900. His paper, later known as "Hilbert's program," was titled "On the future problems of mathematics." Hilbert's program very profoundly influenced the development of mathematics for decades. Reduced to the essentials of his inspiration, the announced goal of that program was to express all mathematics in a formal language. According to that view, mathematics is a purely formal activity with no more inherent significance than a game of chess. When mathematicians make a demonstration, they use axioms and logic.

Hilbert held that simple intuitive practice, employing signs and rules, was subject to serious error. He therefore demanded that demonstrations be analyzed, that their formal workings be objectified, quite apart from their mathematical meaning proper. He dedicated himself enthusiastically to that task of "metamathematics." But in 1931, Kurt Gödel (1906–1978), the young Austrian mathematician—he was only twenty-five years old—suddenly put an end to Hilbert's dream. In a brief paper ("On the formally undecidable propositions of *Principia mathematica* and affinitive systems I"), he established that it is impossible to prove that arithmetic is noncontradictory.

That remarkable text, which resounded like a clap of thunder in the firmament of mathematics, enlisted mathematical methods too complex to be explained here. Let us just consider the major result for which the term "incompletion" is used: Gödel proved mathematically that no system of axioms is complete, that is, to any system of axioms it is possible to add another axiom independent of those of that system and not contradicting them.

As striking as this proof may be, that one must guard against making improper extrapolations to the supposed limits of "scientific knowledge," let alone the irreparable shortcomings of thought. That is philosophizing cheaply and in the worst way possible.

The second revolution that mathematical logic experienced occurred much later, in 1963. It was the work of Paul Cohen, professor of mathematics at Stanford University, an analyst interested in logic who had spent several years with Gödel at Princeton.

Paul Cohen confronted a classical problem which had for centuries tormented minds as eminent as Galileo, Carl Friedrich Gauss and Bernhard Bolzano; is there a cardinal number strictly greater than the countable infinity aleph null of integers and strictly smaller than the cardinal of the continuous infinity of real numbers (two power aleph null equals aleph)? Actually, Georg Canter (1845–1918), creator of the set theory, had proven that aleph is strictly greater than aleph null. But the question that remained was whether an intermediate infinity existed between the integers and the reals. The hypothesis formulated by Cantor, but unproven, that in the set of real numbers any infinite subset is either countable or equipotent to the real numbers was given the name "continuum hypothesis." Every attempt to verify or refute that theory had failed.

Paul Cohen showed that the continuum hypothesis is undecidable: everything depends on the version of the axioms of the set theory adopted. If, for example, I start from plain arithmetic, the axioms of which were formalized by mathematician Giuseppe Peano (1858–1932) of the University of Turin, I can complete that weak system with a family of noncontradictory and independent axioms, for which the continuum theory will be proven; but I can also complete it with axioms for which the theory is false.

These results are extraordinary and all manner of philosophical conclusions is still being drawn from them. But what is their effect on mathematics itself? More generally speaking, what is the impact of the elaborations of logic on the other branches of mathematics?

I shall not take the blunt and even sectarian position of many mathematicians who reply that logic is of no use to them and that it is extraneous to real progress in mathematics. It is, all the same, not immaterial to determine whether a theorem on which a great deal of effort has been spent is decidable or not! Let me add that logic, aside from making possible useful refinements of mathematical language, has, moreover, contributed many new techniques in some areas of calculus. It has, for example, opened the way to so-called nonstandard calculus (where infinitely small or large quantities and finite magnitudes are treated on an equal footing); one should not, however, overestimate its theoretical importance, as was recently the fashion in France.

Furthermore, it cannot be said that logic has ever produced theorems that have turned mathematics itself topsy-turvy. Besides, it is significant that the Fields Medal has never been awarded to a pure logician (of course, Paul Cohen obtained it in 1966, but he was not a pure logician). After all, it does not seem that a great deal of hope should be vested in logic for the future of mathematics, because, in principle, what I would willingly call the "sense of object," the process of delving into that object, is alien to it.

Now, the generalizations to which productive mathematics lead are always guided by the concern to discover and describe new structures offered to research by that sense of object. And logic is, for the most part, manifested by almost mechanical reasoning, formalized to a point focusing on objects axiomatized with well defined rules of calculation. In that sense, it is a precursor of computer science and continues to contribute to its development.

MATHEMATICS AND COMPUTER SCIENCE

Another condition to be observed, if we are to grasp the nature of mathematical abstraction, is not to confuse mathematics and computer science. Yet the social success of computer science and the media prestige of the high priests who promote it only add to that confusion.

I therefore need to say a few words about it, although computer science is not my field of research. Here again, it is wise not to yield to any of those conflicting passions which divide the intellectual community: we should neither scorn the theoretical importance of the work of computer scientists nor rush into giving them medals for scientific creativity. It is a question once again of appreciating its impact on mathematics, considered in all of its inventive aspects, and of charting the right course for that research.

There is no denying that since 1976 computer science has been capable of making an effective contribution to the solution of unresolved mathematical problems. In fact, 1976 is the year when finally the solution was essentially found to one of those most exasperating problems: the so-called four-color problem. Now, it must be admitted that computer science alone enabled it to be solved. That problem, of misleading simplicity, had been posed to the distinguished mathematician Auguste de Morgan by one of his students in 1852: "Is it possible with just four colors (or fewer) to color any map, so that two regions having a common border will never be in the same color?" From William Rowan Hamilton to Heinrich Heesch, countless eminent mathematicians racked their brains over what seemed to be child's play.

I shall not retrace here the turbulent history of the failures that followed one another for over a century. There is irony in it, for everyone believed that each in turn had found the solution before soon realizing that all they had done was to contribute an element of additional complexity to the problem. That complexity was so great that the proof given, which ended up on an IBM 360 computer at the University of Illinois, thanks to the work of computer scientist Jean Koch, is based on the scanning of two thousand maps and required several thousand hours of calculation on that powerful machine. It must be pointed out, however, that it was through mathematical reasoning that the problem was able to be reduced to the study of a finite number of configurations, which were then processed by the computer. If the text of the proof were to be written, a human lifetime would not be long enough to read it!

One may claim that this proof did not open up new fields of research to mathematicians and that the famous theorem remains, and will continue to remain, a simple curiosity in the fabric of mathematics. Perhaps, but it is still a problem of pure mathematics which was thus solved.

Another notable case partially treated in this way: that of the classification of simple finite groups, which required the cooperation of around a hundred people and can be understood only by a high-level expert.

Let us add that computer science clearly has numerous interesting applications, and raises exciting questions, such as the limit of calculability of computers or the design of robot models. It also sheds light on genuinely philosophical questions, bitterly debated for nearly fifty years: will it be possible or not, in principle, to build a machine as "intelligent" as a human? But how then should intelligence be defined? And, assuming it were possible to build such a machine, is it in any sense desirable? Another question, just as important on the practical level, involves the very nature of computer science. In fact, the machines operate on the basis of programs which are, at one stage or another, written by humans—who control their subsequent uses only imperfectly. Now, the slightest typing mistake can have unpredictable consequences, and we are talking of human error—not to mention "viruses" which can freeze machines or erase data in memory. Science fiction has already anticipated machines that escape human control, but the problem is beginning to become real.

The question that interests us for the moment is the following: are we talking about a science that can be given the same epistemological status as mathematics? Or rather is the development of computer science strictly dependent on technical problems? Is it—as is sometimes claimed—just a simple matter of computer technology that tends to usurp the title of science?

It seems to me that no one can dispute the existence of an actual theoretical computer science, even if that theoretician's activity is not the best known or broadest aspect of the work of computer scientists. This theoretical computer science was introduced by Alan Turing (1912–1954). It should never be forgotten that the now famous "Turing machines" correctly presented as prefigurations of our computers originally had no concrete existence and, at first, were purely theoretical products of the questions that ingenious logician pondered regarding the problem of decidability raised by Hilbert. One internationally recognized and respected exponent of that branch of research in France is Marcel-Paul Schutzenberger. But it is undoubtedly still too early to judge the scientific fortunes of this theoretical computer science.

If within the next ten years or so we should witness the development of excellent computer designs capable of solving a growing number of the problems presented and having productive heuristic effects on mathematics and the other sciences, we shall then be able to answer the question affirmatively. There is, after all, no reason to doubt that computer science may, under these conditions, become the great rival of biology in its relations with mathematics as a supplier of problems, aside from physics, of course.

FROM MATHEMATICAL PHYSICS TO PHYSICAL MATHEMATICS

Let us now get back to mathematical abstractions themselves and to the extremely close yet very different relationship they maintain, as has been seen, with the thinking of physicists.

People have too often been content to say that mathematics represents the "language of physics." In a sense, mathematics is undoubtedly a language; but it is primarily thought, thought that in itself is inventive. It is thus an oversimplification to make it into a simple "expression" or the more or less elegant garment of a thought which, if the formula is taken literally, precedes it and remains extraneous to it. Could it be reasonably maintained, for example, that the remarkable feat of mathematical imagination accomplished by Werner Heisenberg in presenting the first complete version of quantum mechanics boils down to a felicitous expression? Of course, not: the mathematical thinking of Heisenberg is so intertwined with his physical thinking that it becomes one with it.

It is also often said that mathematics provides physics with conceptual "tools." It has even become almost a stock phrase. In fact, it is not uncommon for physicists to "help themselves" at the mathematicians' table. The effectiveness of these "tools," developed without any apparent regard for their use, can be seen as wholly enigmatic, as was observed by Nobel Prize winner Eugene P. Wigner who, in a famous statement made in 1959, did not hesitate to speak of a miracle (*"the unreasonable effectiveness of mathematics in the natural sciences"*).

It seems to me that before invoking hypothetical gifts from heaven, it is not unavailing to turn to history. Now, as trivial as that point of view might be judged, it can hardly be disputed, it seems to me, that the abstract mathematical intuitions associated with the imagination of mathematicians borrowed from physical models as a point of departure; it matters little whether the beginnings of this process were set in Egypt or Greece. All things considered, it does not seem surprising to me that mathematics did "attune" to classical physics.

Today, at the tentative end of that process, we start off from what has developed into very highly abstract mathematics, as shaped, transmitted and consolidated in the course of the progressive generalization I mentioned; and these refined mathematical structures, which can assuredly serve as "tools" for physicists, have proven capable of suggesting some fundamental new ideas to us about the concrete world around us, whether that of classical physics or microscopic physics.

Should we talk about a dialectical approach? The fact remains that an overall process of creativity exists, which reflects a kind of coming and going, no aspect of which should be neglected: the fundamentalist physicist discovers an "effect" on the basis of a prior theory; that experimental result will be the subject of the work of a "phenomenological" physicist who will seek out simple rules to account for it, and then will come the theoreticians who will try to elaborate a mechanical model; and, finally, if necessary, a mathematical physicist will say whether that model is compatible or not with the requirements of the general theory to which reference is being made.

If it is indispensable to construct a new theory mathematically, new conclusions may be drawn by calculation from mathematics so enriched, which in turn will suggest other experiments. And the process will be reboosted when a new experimental result comes along and demands a revision of the theory. The history of the hydrogen atom investigation provides an excellent example here: everything began with the discovery of the discrete spectral lines of that atom, which gave rise to a "phenomenological" study; then came theoretician Niels Bohr who elaborated the "quantified" model of the atom's structure. With Heisenberg we witnessed the intervention of a mathematical physicist who, thanks to a development of matrix calculus, truly laid the foundations for the quantum mechanics within the framework of which the work of Bohr finally made sense.

If we overlook that coming and going, if we forget the history leading to our mathematical abstractions, we shall end up becoming entangled in long-standing philosophical questions, perhaps worthy of respect, but without answers, such as: are "mathematical idealities" the pieces of a world with an existence of its own, which the mind discovered step by step, much the way Plato describes the world of Ideas, or are those idealities, produced by the neuron activity of mathematicians, only emanations of the structure of our central nervous system and, in particular, of the brain?

Those questions have been the order of the day for over two millennia: the terms in which they are couched have not changed in substance, even though the progress of mathematics and of biology constantly renews their formulation. A dialogue on this subject was presented in a recent book by Jean-Pierre Changeux and Alain Connes. While the first author emphasizes especially the philosophical context indicated above, the second offers an important contribution to the problem at hand.

Now, then, the inventive power of mathematics is its most striking intrinsic characteristic. Consider, for example, the finite body theory: a whole system of geometry can be developed on its extraordinarily abstract bases. For the moment, that geometry remains purely speculative; although the game is exciting, it has found no concrete application. But who can say that tomorrow might not be different? Let us not forget that Riemannian geometry—named after Georg Friedrich Riemann (1826–1866), professor at the University of Göttingen—was for over half a century in that situation of sublime and gratuitous abstraction, until it found its concrete physical application in Einstein's general relativity theory!

From that point of view, mathematics develops by itself thanks to constructions free and independent of any physical model; it is really a particular way of thinking, a system of unique intense intellectual creativity. And that is why it seems to me once again totally simplistic and erroneous to regard it as a simple tool of physics.

But where does the "boundary" between physics and mathematics lie, one might ask? For my part, I have said that it is blurred at the present time, with the emergence of "physical mathematics." True enough, physical formalisms intrude today into mathematical thinking in order to suggest not only questions, but methods and solutions, and thus solve "purely" mathematical problems. But I do not think the term "boundary" is the most appropriate, for it inevitably evokes the idea of well-delimited territories on the map of knowledge. If there actually is a distinction to be made between the thinking of physicists and that of mathematicians, it is rather in terms of goal to be set. Is a researcher working on "gauge theories" a mathematician or a physicist? Everything depends on the particular goal.

Let us remember that gauge theories are theories of physical fields possessing very great symmetry, dependent on one or more arbitrary functions, deriving, for example, from the possibility of modifying the potential of a field by an arbitrary function without altering the fact that the field responds to a given equation, as in the case of the electromagnetic field. Such theory is at the root of the so-called "electroweak" model which won Steven Weinberg, Abdus Salam and Sheldon Glashow the Nobel Prize in 1979, and it is believed that all unified physical field models will be of that type. The mathematical formulation of those theories calls for ideas which are at the cutting edge of mathematical progress.

That is why I consider it necessary, at the point we have now reached, for everyone to be "bilingual": physicist and mathematician, mathematician and physicist; for everyone to understand both languages; better yet, to speak them. The present evolution of research condemns to ultimate sterility those mathematicians who believe they can confine themselves to the supposedly closed field of "pure" mathematics without any concern for what goes on in physics or other sciences; we can no longer be content, as was still the case even recently, to tell ourselves that, after all, what is being done in mathematics will perhaps some day turn out to be of use when a physicist finds success in it. That is how things may actually happen, but the link between the two disciplines has become so close that we can also hope to profit from the work of physicists in the free endeavor of pure mathematics.

MATHEMATICS AND BIOLOGY

Can it be hoped that other sciences will establish a similar close relationship, if not a union with mathematics? Is mathematics going to grow and further extend its power? That has been the dream of many disciplines for a very long time. Among them biology is certainly the one that has come closest to

65

fulfillment. Statistics made a strong impact on that science, with the famous work of Gregor Mendel on the distribution and hereditary transmission of characters, taking the scientific world of the time so much by surprise that the best minds were unable to grasp its importance. With the advent and then the development of molecular biology, and with the precise knowledge acquired of cellular interactions and structures, a new step forward is being taken. But we should not take shortcuts: we are still far from the situation prevailing in physics. And nothing will ever be gained from hasty, premature and, therefore, superficial mathematization. In fact, as long as biologists are unable to define precisely what has come to be known as a living system, such mathematization, remaining outside the fold, will be more or less mythical.

Those remarks are especially true of the particular living systems endowed with a brain capable of making decisions. Even though we must remain hopeful of such an accommodation between mathematics and biology, there is still good reason today to make a clear-cut distinction between the life sciences and the sciences of the nonliving world. Why is that? It is very obviously because living systems are much more complex than those of the physical world. In spite of the enormous progress achieved, we do not yet have sufficiently detailed knowledge of that complexity for mathematics to take hold and be able to cultivate such knowledge with adequate inventive formalisms.

One point, however, deserves to be stressed here, for it concerns research with an obvious future, involving the relations to be established between biology and quantum mechanics. It is, in fact, inconceivable that the quantum properties of atomic matter do not play a role in the biological system through those "building blocks" of living organisms, the DNA or RNA molecules, similar to what occurs in chemistry.

As far as the brain is concerned, it is known that an important mathematical model (the differential equation of the neuron) is used in current research. The quantum properties of matter must be utilized to understand the structures of the neuronal movements and their interactions. But, conversely, where interpretation of the quantum measurement of external systems—of nonliving matter—is involved, Eugene Wigner suggested, in keeping with Heisenberg's philosophy, that this interpretation should be coupled with the properties of the human brain. An entire branch of research on artificial intelligence consists, for example, in creating neuronal networks capable of form recognitions. Thus, Leon Cooper (1972 Nobel Prize in physics for the superconductivity theory) successfully tackled, by these means, the problem of writing recognition.

Other even more original research paths were taken by Stan Ulam (1909–1984). A former student of the Polish mathematical school that gathered in Lvov around Banach in the 1930s, Ulam immigrated at the end of 1935 to the United States and in 1943 joined the Manhattan Project at Los Alamos, where he collaborated, among others, with Enrico Fermi, invented what he called the Monte Carlo method and became the true father of the hydrogen bomb. Ulam asked the following question: instead of trying to "apply" existing mathematical and physical theories to biology, why not invent new mathematical models that would be better suited to the specific needs of biology? For example, he suggested coming up with new definitions of metrics (distances) in order to take the particular properties of biological molecules into account.

That research is very interesting. It holds out the promise of important developments in years to come.

MATHEMATICS AND ECONOMICS

When collective phenomena are approached, the same hope is manifested just as strongly; but an even firmer reservation has to be expressed. The most instructive example is certainly that of economics, for in this field serious attempts at mathematization have been made, which can be discussed on solid grounds. Now, what have the economists done? Essentially, according to Paul Samuelson, they have borrowed from the laws of thermodynamics, as they govern the evolution of physical systems, and have carried them over to their subjects. That transfer, dubbed "application," enabled them to give definitions of concepts such as "utility function," "capital" and the like. Numerous models thus emerged in both microeconomics and macroeconomics. Those endeavors went much farther than is often believed today. Lagrangian mechanics, that is, classical mechanics in its systematized form, and variation calculations are used very extensively. French mathematician Gérard Debreu of the University of California at Berkeley obtained the Nobel Prize for a mathematical theory of economics and Italian-born mathematician Franco Modigliani of MIT won it for having designed a mathematical model of financial markets. The attempt was also made to apply René Thom's catastrophe theory to "singularity" phenomena in that field. But the question of whether those models have a proven connection with the real remains open, to say the least. Are they not still too tentative?

How can we fail to note, for example, that psychological factors associated with the behavior of economic "agents" compromise the purity and effectiveness of those models? That certainly did not escape the attention of the experts who have tried to apply game theory to that behavior, in order to reduce its disturbing effects. Very evidently, however, game theory, which is a very fine mathematical theory founded in 1944 by John von Neumann and Oskar Morgenstern, will not suffice to perceive all of the parameters, for it has to assume, in principle, that individuals at all times follow the optimal strategies consistent with their interests, which is clearly not the case!

Without any doubt, statistics can then be useful in verifying or confirming certain hypotheses, but it is still too soon for the genuine power of mathematics to take hold in economics, the way it has in physics. The situation is comparable here to what is true of biology: the systems we would like to mathematize are much too complex and entail too many parameters to be properly grasped today. But it is not enough to blame the number of parameters and to set out immediately on a quest for "hidden parameters." For here we have a more serious problem: since we are dealing with systems involving beings endowed with brains and able to adopt a multiplicity of decisions that disturb the way they operate, we do not yet know how to conceptualize them. It is thus essentially because economic theory itself suffers from a lack of conceptualization that the effectiveness of mathematics is for the moment very limited in this field. Will such conceptualization develop in the years to come? We have every reason to hope so. We shall then see whether mathematics, as it exists, will be up to the task, or whether it will be necessary to develop other types of mathematics, other ways of thinking suited to those specific forms of interaction, which would complete and enrich the ones available to us at the present time.

What has just been stated regarding economics could be relevant to all the human and social sciences which "cast longing eyes" at mathematics. The same basic arguments would at the same time show that their hopes are legitimate, but their victory cries are yet premature.

III

BEING A

MATHEMATICIAN

THE YOUTH OF MATHEMATICIANS

It is said of mathematicians, even more than of other scientists, that they can be creative only in their prime of youth as researchers. As we have recently seen, there are some biologists who have tried to support that common opinion with what we know about the evolution of the central nervous system and are beginning to learn about the aging of the brain. In spite of the progress achieved in the past few years and the prospects opened up by developmental neurobiology, however, this knowledge still remains very fragmentary and it does not seem to me that any evidence has been adduced to date for a correlation between that social fact and a biological foundation.

What is certain is that the history of mathematics is full of examples of precocious genius. Evariste Galois, whose brilliant work, which marked a decisive turning point in contemporary mathematics, was tragically and stupidly interrupted when he was twenty, certainly represents the most extraordinary example of that. But such cases, fascinating as they may seem, should not let us forget that examples of the opposite also exist; we have seen mathematicians whose creativity came relatively late in life and still others whose creativity in youth continued into maturity.

Nevertheless, it has been observed statistically that the great discoveries were made or took shape usually before the age of thirty. Perhaps that observation is even truer of mathematics than of physics, contrary to popular belief. Furthermore, according to the rules, the Fields Medal can only be awarded to mathematicians under forty years of age (and just two to four of them are handed out at each International Congress of Mathematicians, which meets every four years).

The biological causes of this phenomenon do not yet seem to me solidly enough established to be invoked, but there can hardly be any doubt, on the other hand, about the intellectual, institutional and social factors contributing to it. If creativity in mathematics, in particular, consists in coming up with new abstract ways of thinking, it presupposes, in effect, a boldness and, let us say, a lack of deference to tradition that are very often lost with age; in this case, not with a dulling of the neurons, but with career advancement. Not to yield to authority, not to bend under the weight of time-honored knowledge: such are the conditions of an inventiveness which demands sustained intellectual effort and buoyancy.

To this must be added that the contemporary organization of science, its institutional and social lifestyle, leads to a situation in which researchers are before long forced to assume administrative and managerial tasks, or else to use or at least promote the research of others rather than pursue their own. Many academic notables thus live off the work of their students and from the theses they direct. They are thus, as it were, lost to research.

How are inventive young mathematicians to be educated and trained? The question is not easy, since, if we follow what has just been said, it essentially involves encouraging rebellious spirits to blossom with free rein to the imagination, preserving a certain nimbleness of mind while affording it the means of being creative. The "training" procedures, as we conceive them and ordinarily practice them, hardly lend themselves, one must admit, to that kind of enticement, since they more often emphasize the transmission of acquired knowledge and apprenticeship in proven methods. And considering that those procedures resemble an obstacle course where the competition is tighter and tighter, this hardly encourages departing from the beaten path. The most extreme example of this is undoubtedly to be found in present-day Japan, the quality of whose mathematical school (though very good) is not consistent with either its extraordinary economic power or its population, partly because of too rigid an educational system.

THE FAULTS OF BOURBAKISM

The question is especially ticklish considering that, under the influence of the French school of mathematics (Bourbaki's), it was wrongly approached for a long time. That school, whose prestige was enormous all over the world, defended, as I have already stated, a formalistic conception of science. It was followed, as was to be expected, by educational practices which had undeniably harmful consequences for several generations of students and researchers. I indicated above that those consequences were not limited to higher education and the training of researchers, but had affected secondary education as well and helped put mathematics in the arbitrary and absurd selective role still assigned to it at present.

Under the influence of Bourbakism, it was wrongly demanded of young people who showed a special talent for mathematics that they concentrate their efforts first of all on growth of knowledge; prodigious reading was required of them; they were trained in the most rigorous thought processes possible. An amazing paradox where scientific research is involved: they were not acknowledged the right to make mistakes or even to approximate the truth! Of course, training in scientific rigor cannot be condemned. But when the approach being taught invariably goes from the most general to the most specific, as was still the case ten or so years ago, the study of mathematics is transformed into a purely talmudic exercise and the liveliest imaginations stagnate. Yet, let me repeat, imagination is what counts in the progress of mathematics; it is the most precious asset. And usually the direction of research runs from the specific to the general.

From that standpoint, the failure of Bourbakism is almost universally recognized today, even in France. The encyclopedic path on which they were led astray for so long is no longer imposed on young researchers. What natural paths should then be open to them? I would unhesitatingly say: the opposite path, the one restoring the process of creativity in mathematics. Instead of initially establishing a general structure and presenting examples only as specific applications of that structure, we should begin with examples that are then studied in depth, as has been done in the United States for quite a while now. Only through work on examples can we attempt to see what general structure might be built to accommodate them.

The advantage of such an "open" process is immediately apparent. Instead of encouraging the selection of talents which are simply good at assimilating knowledge, coordinating and applying it, but without real creative ability, varied talents will be left to express themselves and succeed in the fields to which they are best suited.

These remarks can be illustrated by a relatively simple example. There is, in fact, an area of mathematics that consists of the study of what are called "topological vector spaces." The French method of teaching this, for instance, consisted for a long time in making a complete preliminary tour of the general theory of topological vector space—which is mathematically a rather lean theory—and then studying ever more specific cases, those of the Banach and then the Hilbert spaces. The reverse process consists in starting from what is described as the "Hilbert space," which is a very specific case involving a large number of interesting properties, before proceeding to the general theory. A typical example of Hilbert space is that of all the functions whose square of the absolute value has a finite integral; that space possesses a scalar product defined by the integral of the product of two functions.

Two opposite conceptions of mathematics are actually contrasted here: one showing an idolatrous respect for general structures and venerating their formal beauty; the other finding that the strength and wealth of mathematical thought are manifested in the specific cases, in the most difficult ones. The creative mathematical process begins, like it or not, with examples and progresses with attempts to formulate theorems concerning them.

It is understandable how, under these conditions, mathematics can be the object of genuine passion on the part of its devotees. That passion, which is often surprising to the uninitiated and sometimes provokes their irony, if not their sarcasm, unhesitatingly expresses its motives in aesthetic terms: who has not heard a mathematician speaking of a "beautiful theorem" or of an "elegant" proof? Much reasonably sound philosophic speculation has been given over to these expressions. Now, even if I sometimes abandon myself to the use of such vocabulary, I am not sure it is very propitious. We should seriously wonder about what we call a "beautiful" theorem before rushing ahead and making comparisons with art, in order, finally, to celebrate their union philosophically in the element of a quality ("beauty") that is proclaimed to be universal!

Let me recall a fact learned from personal experience. I happened to be teaching for one year at the University of Kyoto; I never heard any Japanese mathematician using the kind of aesthetic vocabulary that is commonplace among us. That small fact warrants, it seems to me, an hypothesis. When we speak of a "beautiful" theorem, we mean that it fits particularly well into the traditions of mathematical thought to which we are referring; in short, it recalls to us something of that tradition. If we describe that harmonious recollection as "beautiful," is it not by virtue of the fundamentally Greek conception of beauty we have inherited? Under those conditions, should we not suspect such vocabulary, which emphasizes the formal aspect of mathematical creation, of being more conservative than is appropriate for a discipline that makes progress only by drastically altering its content?

On the other hand, though, mathematicians all over the world (probably influenced by the French school of mathematics) will talk about "trivial" notions or results, about "trivialities," to describe facts which seem so evident to them that they become trite. And this expression, which does not at all have the rough connotation usually attached to triviality in the literary sense, is sometimes disconcerting to nonmathematicians: it could almost be described as shop talk.

MATHEMATICS AND CULTURES

Regardless of these difficult questions, we must emphasize the bond that indeed exists between the existence of mathematical research proper and a given social and cultural milieu. If it is right to highlight the free creation that guides the progress of mathematics, it should not therefore be concluded that this activity is radically cut off from the world in which it is deployed.

Actually, history has furnished eloquent proof since antiquity that great mathematicians do not appear just anywhere or haphazardly. It implies an extraordinary cultural preparation on the part of a given society; sociology as a whole is involved, a broad effort which must mobilize considerable spiritual and material forces. And anyone who remembers that mathematical creation is a highly intellectual activity will not be surprised to find that it develops in societies where artistic creation has attained a considerable degree of refinement. The "golden ages" of mathematics over the centuries developed in the course of golden ages of civilizations (the converse not always being true).

It is understandable why, under these conditions, although mathematics requires no heavy equipment and no particularly costly investments, the Third World countries are not producing great mathematicians, with a few rare and remarkable exceptions, like Brazil or Argentina (but those are countries of immigration). That situation is also true of the Arab countries which, though the creators of algebra, today have very few important mathematicians.

In fact, until the 19th century, most mathematicians have been French, German, or Russian (along with some remarkable individuals in other European countries). Japan learned how to establish, by imitation and by methods suited to its circumstances, the conditions appropriate for the training of a high-level school of mathematics. At present there are even approximately one million amateur mathematicians in Japan (which has a population double that of France and nearly half that of the United States)! The reservations we expressed above are no less disturbing with regard to the future development on the highest level of the Japanese school, owing to the harmful effect on the creative mind of an extremely finicky selection at every stage of education.

As for the United States, it was not really able to develop its mathematical research until after the Second World War, thanks to large-scale immigration of European and some Asian mathematicians, in furtherance of a determined policy of "importing" brain power. It should be added, however, that its motives were—and on the whole remain—primarily economic. Its interest in mathematicians is a function of the assessment it makes of the economic importance of their work. In general, it is not out of love for scholars that it fosters research, but with a view to anticipated technological success. That has been seen with the rise and fall of credits allocated to mathematicians by the National Science Foundation, which coordinates the bulk of federal aid to mathematics research.

That situation is not without risk or detriment to the social status of researchers, who are infinitely less highly regarded in the United States than administrators, executives and engineers of every kind.

One question suggested by the foregoing remarks is often asked with various more or less honest mental reservations and ought not be ignored: why has there been a great abundance of Jewish physicists and mathematicians? Quite apart from theological and political considerations, which lead to fanaticism, or genetic considerations, which feed racism and are, moreover, without real foundation, it will be seen that this fact can corroborate the cultural approach I am defending. How, in fact, can we not relate the existence of a culture of the Book, which has for centuries been inculcated in all Jewish children, and in such a way as to develop critical reasoning and inquiry, to their attraction to anything intellectual and anything written?

One may say that other religions of the Book exist, too. Undoubtedly, but the contrast is very sharp between this tradition and, for example, that of the Catholic Church, which has for a long time—a very long time—fought science and always suspected it, more or less openly, of representing a threat to the faith. Jewish tradition, for its part, never placed itself in conflict with the great scholars. It has never experienced a Galileo affair; and

most rabbis have always very cleverly accommo-
dated themselves to the discoveries of science, by
unhesitatingly revising their symbolistic readings
and interpretations of the sacred texts. Furthermore,
in contrast to other religious traditions, and even in
total opposition to the Islamic tradition (which does
not tolerate the slightest interpretation of the written
text), Judaism has always fostered and even broad-
ened not only study of the Scriptures, but also
commentary, not to mention the development of
treasures of the imagination in the search (in the fin-
est Talmudic tradition) for the best ways of taking
liberties with the laws of the Lord.

To which one must finally no doubt add the
"minority spirit" and the quota system which have
pushed many Jews of the Diaspora into the most
advanced studies, in which they have proven their
excellence in an effort to achieve social protection
and upward mobility. The *a contrario* proof of this
argument could be drawn from the history of the
State of Israel, where that minority spirit no longer
applies. Israel embraces approximately twenty-
seven percent of the world's Jews. Nearly one-third
of the Nobel laureates in physics during the past
half-century have been Jews. Not one of them is
Israeli! Not one Fields Medal has been awarded to
an Israeli!

POLITICS OF RESEARCH

These remarks are, however, still insufficient to understand the mathematician's social being. It is necessary to add considerations which touch upon the politics of research and on politics *per se*.

As far as the politics of research is concerned, the question has taken a sharp turn because of the rising force of computer science. The current trend in research raises questions which concern computer scientists as much as mathematicians. As I have stated, a theoretical computer science exists, on which some hope for the future of mathematics in general can be predicated. But while this type of research requires only a limited number of researchers, we are witnessing a genuine stampede of mathematicians—often second-rate—into this field. Notwithstanding the important economic and social role being played by computer science, we must prevent the establishment of an imbalance that could very soon prove unhealthy to computer science itself. The media and the politicians will very soon have to realize that this type of artificial growth is cancerous. Specifically, can students be allowed to rush into computer science studies, when it is known for a fact that the market is becoming saturated—at least as far as technical experts are concerned?

It is true—and here general politics is involved—that this mad rush also has financial reasons: what student would be willing to risk embarking on lengthy studies in order to share the lot, if successful, of a university professor dedicated to research and teaching in our field at a salary far below what industry offers a good computer scientist? The university recruiting crisis raging in France, for instance, is well known. It is very serious. In varying degrees, the problem is international. It is a matter of urgency that university authorities and government officials should take responsibility.

Let us envisage the problem from its most general aspect: in every country where the organization of research and higher education is centralized, a politics of need thus evolves without taking into account the means available. When, owing to international competition, technological development and the fact that the effects were not known or could not be foreseen in time, a need suddenly crops up (as in computer science, electronics or telecommunications), we face an almost "catastrophic" situation, in the sense described in the work of René Thom: an "infinite" demand is answered by a "zero" supply! For lack of preparation, there are not enough qualified people available to engage in research or teaching in what is now proclaimed a priority area.

What then are the politicians, the higher civil servants and other administrators, who often interchangeably hold the decision-making positions, doing about it? They unreasonably increase the number of positions in the areas considered! And suddenly the universities are summoned to fill those positions as quickly as possible. What do the faculty members do? For fear of losing their job funding (a situation well known in university circles), they hasten to recruit on the run the first comer, lowering the hurdle as much as it takes! Thus, some young people, who often have pretty much failed in their original pursuits, but have had contacts with the new field, will be barring the way for any legitimate candidates who might come along for the next thirty years.

To complete the picture, we should add that, in such cases, with a competing job market offering better financial conditions, the most competent will not even apply for these university positions.

Another way to proceed should be substituted for this disastrous logic: let the universities be confronted with their responsibilities; let any success in hiring be rewarded *a posteriori* with State aid; and let any failure be penalized by withdrawal or reduction of that aid. This implies that academia will be given more leeway, while instituting very strict and vigilant scientific control. The authorities would then no longer be content, as they are today, to present a policy which looks fine on paper and satisfies only the administrators who, in France, unfortunately have almost no genuine contact with the realities of research.

THE TEACHING OF MATHEMATICS: MODERN MATH

In the late 1960s André Lichnerowicz, an eminent mathematician, was appointed to preside over a commission that was to organize a reform of the mathematics programs of French secondary schools. It was a matter of adapting education in this field to the requirements of the modern world, to research in mathematics as well as to the living mathematics being used in physics and the other sciences. "Modern math" was going to be introduced in the educational institutions! An initiative of the same type was taken at the same time in Belgium (Monsieur Papy was "Mr. Modern Math" there); shortly afterwards a similar experiment was undertaken in the United States under the banner of the "*new math.*"

The basic idea was excellent, but in France the realization was disastrous. A kind of national psychodrama ensued, which it is not useless to reflect upon if we are to avoid a repetition of such mistakes in the future. The first mistake was to cut mathematics off from its intuitive foundations and thus render it more abstract for children at a tender age. That error was immediately criticized by many mathematicians and, in particular, by Jean Leray, a great expert on differential equations. Rather than cutting mathematics off from its intuitive foundations, what should have been done, on progressing into abstraction and axiomatization, was to enrich that base and multiply the examples with which pupils could maintain a relationship of familiarity. Although André Lichnerowicz was not himself a Bourbakist, one might say that on that occasion a kind of "Bourbakism of the secondary schools" triumphed, much to the distress of the students (and teachers)!

The second mistake provides a general lesson about reforms decided by the political powers. When such a decision is made, it is certainly necessary to know what is desirable, but is also advisable to take into account the means available. Now, in this case, those means consisted of the teaching profession, as it then existed—and which had not been trained in "modern math"! Many teachers were not ready to teach the new mathematics; they had no sound knowledge of the subject. Failing to understand, they were unable to show their pupils how a given abstraction was linked to living examples. The result was that those teachers taught it as a set of abstract expressions that their students had to learn by heart, literally, without regard to the very meaning of the proofs and theorems! Abstraction was sometimes pushed to a point where terminology was used that confounded even professional mathematicians. (A number of them even received bad grades in their children's homework for not having used the exact formulation desired by the district education inspectors.)

It took years for those two errors to be essentially corrected. That has now been accomplished for the most part, even though teaching still remains often too abstract and unilaterally emphasizes the axiomatic process and not the fruitful investigation of concrete examples.

CASINOS

Let me end this small volume on a lighter note, without thereby departing from the very serious question of the power of mathematics. The reader will immediately make the connection with what has preceded.

Mathematics has for ages maintained a close relationship with games of chance. We need only think of Pascal! And there is no shortage of gamblers, professional or otherwise, who believe that this science could provide a "good method" of winning at the gaming tables, since, after all, it is always a question of probabilities. Let us see how true this is.

We shall begin with the best known system, the martingale method. Let us make one thing clear: in mathematical terms, martingales do not stand the test. What is more, they are extremely risky in practice. Let us take a classic situation at roulette, where there are thirty-six numbers plus zero: playing equal chances (black and red, odd and even, manque and passe), that is, betting on eighteen numbers, you place a wager of x, and then $2x, 4x, 8x \ldots$ If, after having bet x, you lose, but win on the following spin, you will have won a total of $2x - 1x = x$. Now, if you lose x and then $2x$ and win $4x$, you will have still taken in one time x. This is therefore claimed to be an infallible method for winning. Let us assume that $x = 1,000$ dollars; it is contended that, having risked that sum at the outset, you will have doubled it, whatever happens, after a number of lucky spins. But it should be noted straight away that luck can take some time coming and that, with this method of gambling, one has to have an extraordinary amount of capital available to be able to "wait." Repeat the calculations with 1,000 dollars. Should you lose four times in succession and win only on the fifth spin, which is not at all unusual, you would require a capital of at least 31,000 dollars in order to win 1,000 dollars in the end. I leave it up to you to calculate the sum that would be needed if the series were to consist of twenty spins!

Let us note, finally, that if you do not win on the fifth spin, you will have to stake 32,000 dollars on the next one and double that amount on each additional spin. Now, that is impossible, because the bets are limited. As every casino in the world thus sets ceilings, I have put it mildly when I said that the method is risky. For similar reasons, no martingale, however sophisticated, can beat the house.

The only worthwhile "method" is quite different. It is based on the fact that no roulette wheel is perfect nor is any croupier. (I am, of course, not referring to a dishonest croupier!) The situation is very different here from that of lotto. With great training, it is then possible to study for free the correlations emerging from the fact that the croupier grows tired and will inevitably start throwing the ball with a certain degree of regularity. It will then also have to be taken into account that the roulette wheel itself, which does not change, when the croupier is relieved, establishes, due to its inevitable imperfections, some probabilities that are stronger than others.

Correlations can thus be discovered on different numbers: one may know, for example, once the 6 comes out, what neighborhood is going to be favored on the following spin. If I discover that, in one spin out of three, one-sixth of the wheel is thus favored, out of thirty-six numbers—if I am quick about it, for the numbers are not arranged on the baize in the same order they occupy on the roulette wheel—all I have to do is play six or seven numbers to win one spin out of three, which in the end makes the game profitable.

What is involved here is neither a martingale nor pure mathematics, but rather a study of the defects of the roulette wheel and of the croupier and a calculation of correlations.

How can we fail to see in the feats of certain gamblers who thus break the bank (sometimes aided by computers) a marvelous demonstration of the power of mathematics? I now leave the reader to dream, for mathematics, so understood, also makes dreams possible.

BIBLIOGRAPHY

BOURBAKI, N.: *Eléments de mathématique* [Elementary mathematics] (in particular, Book I, "Théorie des ensembles" [Set theory], Hermann, Paris.

EINSTEIN, A.: *Relativity (The Special and General Theory)*, Crown Publishers, New York, 1961.

FEYNMAN, R.: *The Character of the Physical Law*, MIT Press, 1967.

GINDIKIN, S.G.: *Tales of Physicists and Mathematicians*, Birkhäuser Boston, 1987.

HADAMARD, J.: *The Psychology of Invention in the Mathematical Field*, Princeton University Press, 1945; Dover, New York, 1954.

HEISENBERG, W.: *Physics and Beyond*, Harper & Row, New York, 1971.

PAIS, A.: *Inward Bound (Of Matter and Forces in the Physical World)*, Oxford University Press, 1988.

SHUBNIKOV, A.V., and V.A. KOPTSIK: *Symmetry in Science and Art*, Plenum Press, 1974.

ULAM, S.: *Science, Computers and People (from the Tree of Mathematics)*, Birkhäuser, Boston, 1986.

WIGNER, E.P.: *Symmetries and Reflections (Scientific Essays)*, MIT Press, 1967.